SET THEORY AND LINGUISTICS

JANUA LINGUARUM

STUDIA MEMORIAE
NICOLAI VAN WIJK DEDICATA

edenda curat

C. H. VAN SCHOONEVELD

INDIANA UNIVERSITY

SERIES MINOR

NR. 70

1968
MOUTON
THE HAGUE · PARIS

SET THEORY AND LINGUISTICS

by

ALEJANDRO ORTIZ and ERNESTO ZIERER
UNIVERSIDAD NACIONAL DE TRUJILLO

1968
MOUTON
THE HAGUE · PARIS

English version: Richard Hawkes

LIBRARY OF CONGRESS CATALOG CARD NUMBER 68-17900

Printed in The Netherlands by Mouton & Co., Printers, The Hague.

FOREWORD

We are said to be living in the cybernetic age. Synthesis is the approach advocated to reverse the tend towards increasing specialisation, a sort of bridge-building between different academic disciplines. This new concept was one of the two reasons which induced us to organise an interdisciplinary seminar, the result of which forms the subject-matter of this book.[1] Our second stimulus was the fact that on the one hand, very few linguists have any major mathematical training, and on the other, the use of mathematical methods in linguistic research is increasing every day.

In talking about the application of mathematical methods to a humanistic discipline par excellence like the science of language, we also bore in mind the following points:

a. Mathematics works with clearly defined and verifiable concepts.

b. Mathematical notation is concise and precise.

Linguistics can make use of these advantages to examine different grammatical theories, formulate them more accurately, and discover new aspects of them not seen previously through lack of a methodical system of linguistic research.

We should however give a warning that in linguistics the only function of mathematical methods is to describe linguistic phenomena more clearly; they are merely convenient terms of description. We must also make clear that not all linguistic problems can be successfully dealt with mathematically. This is the main reason why we do not agree with the term 'mathematical linguistics'.

In this seminar we discussed the application of set theory to

[1] The seminar was held in 1965 in the National University of Trujillo, Peru, and organized by the Department of Foreign Languages and Linguistics.

linguistics. The seminar was organised as follows: Each study session was divided into two parts; one of them was devoted to the explanation of the mathematical theory involved, and the other to its application to linguistic phenomena. Both parts have been incorporated into this publication, whose sole purpose is to awaken the reader's interest in the potential importance of modern algebra in linguistics.

For the reader interested in further applications of modern algebra to linguistics we have included a selected bibliography.

Trujillo, Peru, July 1966
National University of Trujillo

ALEJANDRO ORTIZ
ERNESTO ZIERER

CONTENTS

I

SETS AND ELEMENTS

1. THE CONCEPT OF A SET

In the course of discussion a definition different from those used in mathematical texts was formulated. We defined the term 'set' as the criterion by which we intend to classify certain objects. This definition gives us a better understanding of an empty set, as we shall see later.

The individual objects which belong to a set are called *elements*. If element *a* belongs to set *A*, now that we understand by set the very collection or aggregate of elements, this will be written $a \in A$. The sign $\in$ means "belongs to". If *a* does not belong to set *A* we write $a \notin A$.

Example 1: Instead of making an a priori definition of a linguistic category, for example saying that an adjective is a word which expresses a quality of a noun, etc., we shall define adjectives structurally as follows:

Let all the words which fit the blank of pattern (1) be elements of set *A*.

"Article + ________ + noun + predicate." (1)

As we can see, only words which tell us something about the noun may fill the blank. We may apply the term 'adjective' to these words. A set may be represented by a diagram.[1]

[1] Such a diagram is called a 'Venn diagram', after the English logician John Venn (1834-1883).

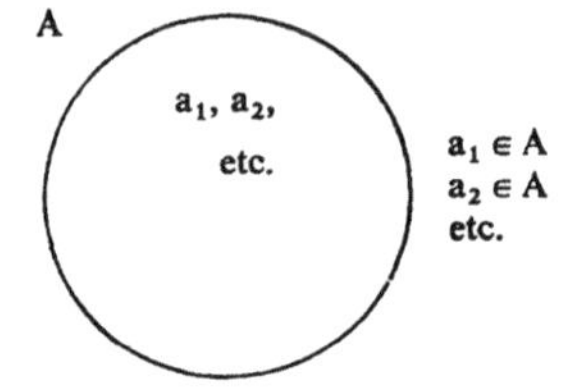

2. THE EMPTY SET

Cases may arise in which we cannot find elements to place in a set which we intend to form according to a pre-fixed criterion.

Example 2: Let all the words which fit the blank of pattern (2) belong to set *B*. (Pattern (2) is in Spanish).

"Ha ________ comprado un libro." (2)

In accordance with the grammatical rules governing the Spanish language it is not possible to put a word between the auxiliary verb and the main verb. Thus set *B* which we intended to form is *empty*: $B = \emptyset$

3. FINITE AND INFINITE SETS — DESCRIPTION OF SETS

Example 3: If we classify by making all the sentences based on pattern (3) belong to the same set, we should have to admit that this set would in practice be an almost infinite number of elements.

"Subject + Predicate." (3)

We say 'in practice' although it could be argued that the number of sentences that can be formed is limited by the vocabulary of language L, but it would be impossible, at least in practice, to enumerate all the sentences in language L based on pattern (3). The set is *infinite*.

On the other hand, if we form a set of elements which fit the blank of pattern (4), we immediately note that such a set must be finite.

"I am trying _______ learn Japanese." (4)

Since the verb 'try' must be followed by the particle 'to', the set consists of only one element, the particle 'to', i.e. it is a *finite* set.

The following notation is also used to describe sets:

$$A = \{a \mid a \text{ is an adjective}\}$$

This expression means: "The set of all the elements *a* such that *a* is an adjective". The sign | means "such that".

4. SUBSETS — THE UNIVERSE — EQUAL SETS

We say that set *B* is a subset of set *A* if every element of *B* is an element of *A*. This is denoted thus: $B \subset A$.

We shall express this definition as follows:

$$B \subset A \text{ if } \forall x \in B \rightarrow x \in A$$

$\forall$ means "for every"; this symbol is also called the "alloperator"; $\rightarrow$ means "implies".

Example 4: Let *A* be the set of all the elements *a* which fit the blank of pattern (1), and B the set of all the elements *b* which fit the blank of pattern (5), *B* will be a subset of *A* since there will be some adjectives which cannot refer to human beings, for example 'printed' or 'cuneiform'.

"Article + _____ + noun (human being) + predicate." (5)

This may also be illustrated by the following diagram:

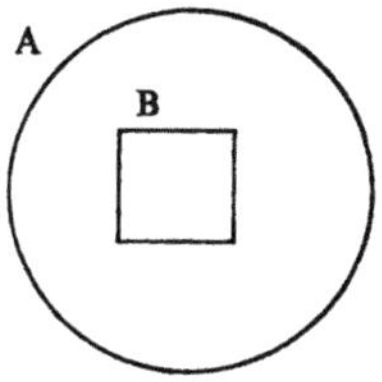

Instead of writing $B \subset A$ we may also write $A \supset B$.

Cases may also arise in which all the elements of a set are also elements of a subset. In this case we write B $\subseteq$ A or A $\supseteq$ B.

Example 5: Let *A* be the set of all the elements which grammatically express the third person singular which fit pattern (6), and *B* the set of all the elements which grammatically express the third person singular which fit pattern (7). The *B* is a subset of *A*, but *A* is also a subset of *B*, since any element which fits pattern (6) can also fit pattern (7).

"________ is etc." (6)

"________ was etc." (7)

We say that set *B* is a *proper subset* of set *A* if and only if *B* is a subset of *A* and at least one element of *A* is not an element of *B*. We then write B $\subset$ A. This is the case in example 4.

In order to limit the discussion of any problem in terms of sets, we start from the *universal set*, or *universe*, symbolised as *U*. The universe in the examples given would be the vocabulary of the English language.

A set *A* is equal to a set *B* if both have the same elements, that is to say if each element which belongs to *A* also belongs to *B*, and if each element which belongs to *B* also belongs to *A*. Equality of sets is denoted thus: A $=$ B. In other words, A $\subset$ B and B $\subset$ A.

5. OPERATIONS ON SETS

Let *A* and *B* be two subsets of set *U* (which is the universe and is the set within which we shall form other sets); then we can form other subsets on the model of *A* and *B* by means of the following operations:

a. *Union of Sets*

The union of set *A* and set *B*, written A $\cup$ B, is the set of elements which belong to *A* or to *B* or to both. This is written as follows:

$$A \cup B = \{x \mid x \in A \ V x \in B\}$$

The symbol V means 'or' in the sense of 'vel' (Latin inclusive or).

Example 6: Let *A* be the set of all the adjectives applicable to human beings and *B* the set of all the adjectives applicable to things or animals (including abstract nouns); the union of these two sets will be a new set composed of all adjectives.

A = {amiable, polite, great...}
B = {easy, tubular, great...}

When union is effected between these two sets any element which appears in both, in our example 'great', need only be mentioned once.

C = {amiable, polite, easy, tubular, great...}

b. *Intersection of Sets*

The intersection of *A* and *B*, denoted by $A \cap B$, is the set of elements which belong simultaneously to *A* and to *B*. We write:

$$A \cap B = \{x \mid x \in A \text{ and } x \in B\}$$

The intersection of the two sets *A* and *B* may be illustrated by the following Venn diagram:

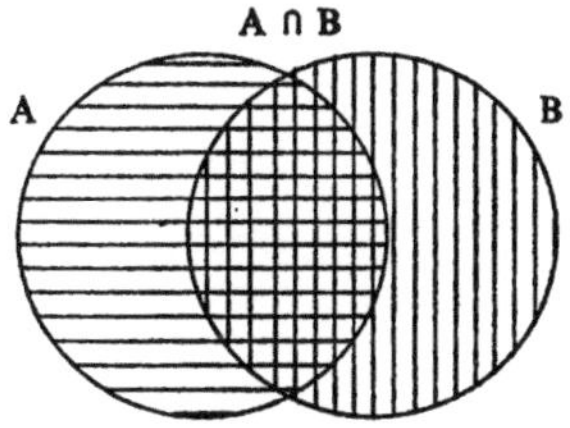

c. *Complement of a Set*

The complement of set *A*, denoted *A'*, is the set of all the elements of the universe *U* which are not elements of *A*. We write:

$$A' = \{x \mid x \in U \text{ and } x \notin A\}$$

Example 7: Let *A* be the set of all the elements which fit pattern (8), and *B* the set of all the elements which fit pattern (9); it would be interesting to know if there are any elements which fit both in pattern (8) and in pattern (9), that is to say whether the two sets *A* and *B* form an intersection *C*. In fact such an intersection does exist, since the word 'well' is sometimes an adverb and sometimes a noun.

"subject + verb + ________" (8)
"subject + verb + article + ________" (9)

"They swim well".
"They found the well".

The intersection of this example may be illustrated by the following Venn diagram. In it the complement of *A* (with reference to the universe *U* represented by the rectangle) is the area marked with horizontal lines plus the area marked with vertical lines, the latter of which is the intersection, where the word 'well' would have to be placed, together with other words which have the same functional characteristics as 'well'.

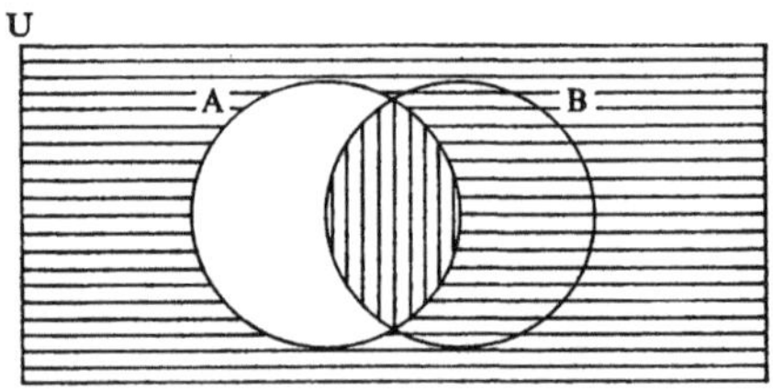

Example 8: People studying a foreign language soon realize that very often the meaning of a word in the foreign language does not correspond exactly with the meaning in their mother tongue. Thus the adjective 'ancho' in the following Spanish sentences has two equivalents in German, depending on whether we are talking about width in volume, or the distance between two points.

Spanish	*German*
'Mi saco es ancho'	'Mein Rock ist *weit*'
'La calle es ancha'	'Die Strasze ist *breit*'

If *A* is the set of Spanish nouns which can take the adjective 'ancho' referring to volume and/or the distance between two points, *B* the set of German nouns which require the adjective 'weit' when referring to width in volume. and *C* the set of German nouns which require the adjective 'breit' to refer to the distance between two points, sets *B* and *C* will be subsets of *A*.

Example 9: Another interesting case concerns what are known as cognates. For example the German word 'Kontrolle' does not coincide exactly with the corresponding word in English 'control'. Let us take the following sentences:

German	*English*
'unter der *Kontrolle* von'	'under the *control* of'
'die Lage *beherrschen*'	'to have *control* over the situation'
'*Kontrollpunkt* Charlie'	'*Checkpoint* Charlie' (in Berlin)

If *A* is the set of the *signifiées* whose *signifiants* are 'control' and 'Kontrolle', *B* the set of the *signifiées* which in English have the *signifiant* 'control' and in German have a *signifiant* other than 'Kontrolle', and *C* the set of the *signifiées* which in German have the *signifiant* 'Kontrolle' and in English have a *signifiant* other than 'control', set *C* is the intersection of sets *A* and *B*.

Example 10: Lexie Content[2]*:* We can define the word 'lexie' as follows: If a word — in Bloomfield's definition: a minimum free form — together with another or other words does not constitute a new meaning distinct from that of each of them, it is called a *lexie*. For example: 'mesa de madera' = 3 words, but 2 lexies. The consequence of this definition is that several words can make up one lexie when their meaning together is different from that of the

[2] The term was introduced by B. Pottier, *Introduction à l'Étude des Structures Grammaticales Fondamentales* (Nancy, 1964), pp. 1-2.

individual words. For example in Spanish 'caballo de vapor' = 3 words, but only 1 lexie.

The lexico-semantic content is called the *sememe*, which is in its turn made up of minimum units of meaning, called *semes*.[3] The following diagram shows the composition of the words 'cup' and 'glass'.

semes	sememe	
	"cup"	"glass"
s_1: usually for containing something temporarily	+	–
s_2: usually with handle	+	–
s_3: usually made of chinaware	+	–
s_4: usually made of glass	–	+

The sememe corresponding to each of the lexies is the set of each one's semes. The intersection of the two sets is called the *archisememe A*.

$$S_1 \cap S_2 = A = \{s_1\}$$

The archisememe is then the set of the characteristics common to both sememes, which in our example is the seme (s_1). This archisememe corresponds to the meaning of the word 'receptacle'. The archisememe is a subset of each of the sets:

$$A \subset S_1, A \subset S_2$$

We can define the lexie more exactly, particularly when it forms part of a syntactical construction.

The content of a lexie will be set *L* made up of the following subsets.[4]

1) Subset *S*, called a sememe, with minimum meanings, called

[3] See: B. Pottier, *Recherches sur l'Analyse Sémantique en Linguistique et en Traduction Automatique*, Série A, *Linguistique Appliquée et Traduction Automatique II* (Université de Nancy, 1963).
[4] See: E. Zierer: 'Sobre el Contenido de los Lexos', paper presented to the IInd Interamerican Congress of Linguistics, Philology, and Language Teaching, Montevideo, January 1966.

semes, as elements. For example the Spanish verb 'alzar', which we can define as follows: 'to move a thing upwards'. We arrange the semes contained in this definition as follows:

s_1 = to move, s_2 = a thing, s_3 = from below,
s_4 = upwards
$S = \{s_1, s_2, s_3, s_4\}$

2) Subset *G*, made up of syntactical requirements of the lexie concerning other elements of the linguistic utterance, and which we shall call the *exigeme*. In this sense the word 'alzar' may be defined as a word which requires an agent and whose action is directed towards a goal; it is transitive.

$G = \{g, tr\}$

Subset *E*, called an *episeme*, in its turn made up of the following two subsets:
a) Subset *C*, called a *conceptueme*, which in our example indicates that the word is conceived in terms of action; thus we must consider the factors of time (t), aspect (a), and voice (v):

$C = \{t, a, v\}$

b) Subset *T*, called a *categoreme*, which tells us which linguistic category we are dealing with (noun, verb, etc.). In our example the word is formulated as a verb, which implies that we must consider the factors of number (n), person (p), and mood (m):

$T = \{n, p, m\}$

The content of the lexie 'alzar' would be formulated by the following set:

$L = \{S, G, E\} = \{S, G \{C, T\}\}$

It is a set of sets.

d. *Difference Between Sets*

The difference between sets *A* and *B* is the set of elements which belong to *A*, but do not belong to *B*. We shall denote this difference: $A \sim B$. Returning to example 7 we have the following sets:

1) Set A of all the elements which fit pattern (8).
2) Set B of all the elements which fit pattern (9).
3) Set C, the intersection of $A \cap B$, of all the elements which fit both pattern (8) and pattern (9).
4) Set D, the difference $A \sim B$, of those elements which fit pattern (8), but not pattern (9), for example 'easily'.
5) Set E, the difference $B \sim A$, of those elements which fit pattern (9), but not pattern (8), for example 'party'.

In the following diagram sets D and E are indicated by the areas with horizontal and vertical shading respectively.

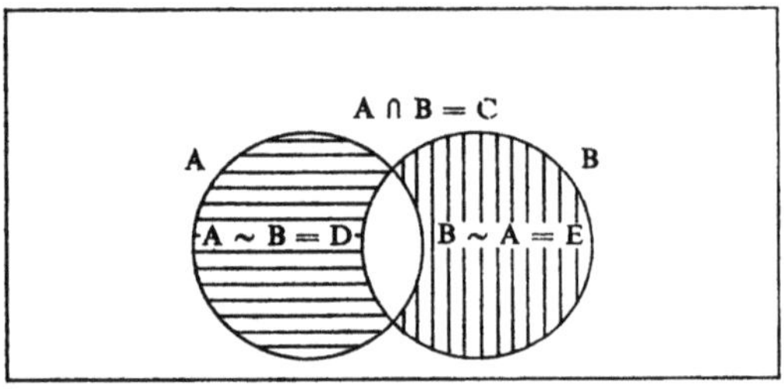

Example 11: We frequently come across expressions which when carefully examined are seen to be ambiguous, that is to say various interpretations of them are possible. To eliminate these ambiguities we must use more complicated syntactical structures. But there are also sentences in which the ambiguities can only be eliminated by the formulation of a series of explanatory sentences, or by intonation. In such cases the formulation of the semantic content in terms of sets, subsets and intersections is a very suitable technique of analysis to el minate ambiguities, as we shall show in this example.

Let us suppose that the following English phrase has to be translated into Spanish, taking into account the different interpretations possible:

"My small car imported from Japan."

Now let us break down the sentence into its *immediate constituents*, limiting ourselves in this analysis to the following four symbols:[5]

n ... nominal nucleus: 'car'
a_1 ... attribute 1: 'my'
a_2 ... attribute 2: 'small'
a_3 ... attribute 3: 'imported from Japan'

Interpretation 1:

a) There are cars imported from Japan.
b) Among these cars there are also some small ones.
c) Among these latter there is one which is mine.

This interpretation can be formulated in terms of set theory as follows:

$$I(1) : n \cap a_3 \cap a_2 \cap a_1$$

In the following model, which represents a derivation tree according to the theory of immediate constituents, the arrows indicate the direction of the determination of the elements, and the numbers the sequence. We can also notice the intersections of the four sets.

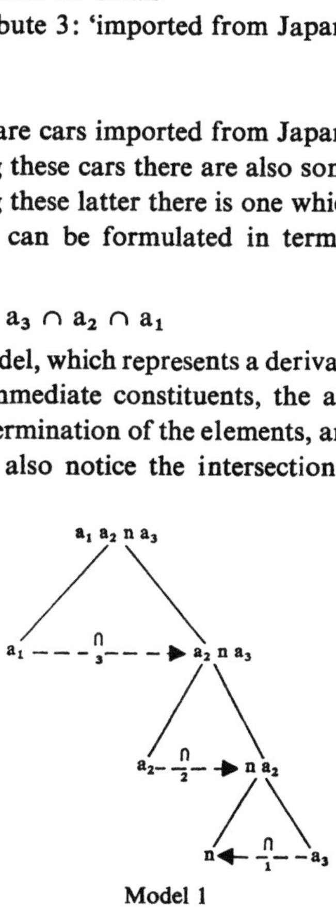

Model 1

[5] On the theory of immediate constituents see: E. Bach, *An Introduction to Transformational Grammars* (New York, 1964); N. Chomsky, *Syntactic Structures* (The Hague, 1957); H. A. Gleason, *An Introduction to Descriptive Linguistics* (New York, 1961); P. Postal, *Constituent Structure*, Supplement to *International Journal of American Linguistics* (1964); R. S. Wells, "Immediate Constituents", *Language*, 23, pp. 81-117, 1947. E. Zierer, *El Orden Jerárquico en las Estructuras Sintácticas* (Trujillo, 1961).

Interpretation 2:

a) There are cars imported from Japan.
b) Several of them are mine.
c) Among the latter only one is small.

This interpretation may be formulated in terms of set theory as follows:

$$I(2) : n \cap a_3 \cap a_1 \subset a_2$$

Model 2 represents the derivation tree for this interpretation.

Interpretation 3:

a) There are small cars.
b) Among them there are some which are mine.
c) Among the latter there is one imported from Japan.

This interpretation may be formulated in terms of set theory as follows:

$$I(3) : n \cap a_2 \cap a_1 \subset a_3$$

Model 3 represents the derivation tree for this interpretation.

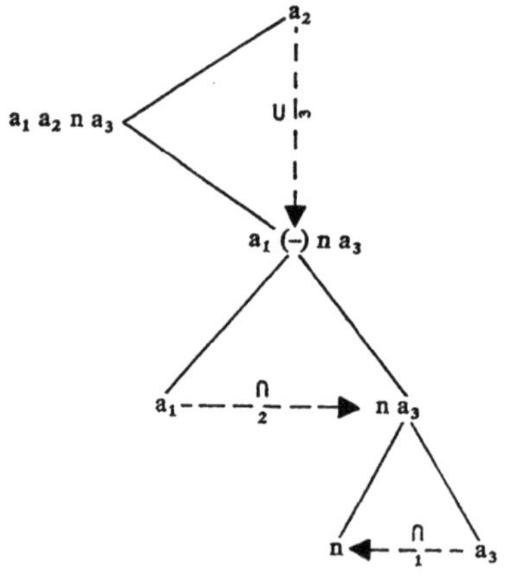

Model 2[6]

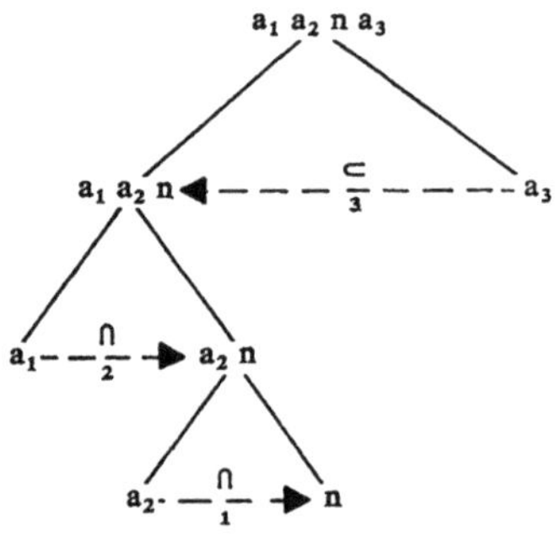

Model 3

[6] The symbols $\cap$, $\subset$ should be read from the same direction as the numbers under them.

Interpretation 4:

a) Several cars are mine.

b) Among them there are several small ones.

c) Among the latter there is one imported from Japan.

This interpretation may be formulated in terms of set theory as follows:

$$I(4) : n \cap a_1 \cap a_2 \subset a_3$$

Model 4 represents the derivation tree for this interpretation.

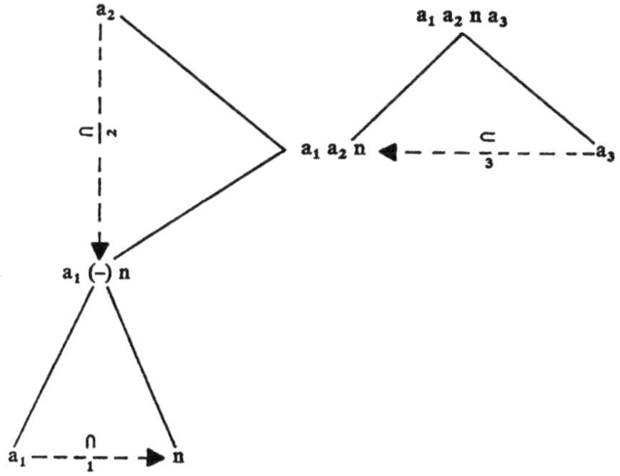

Model 4

Interpretation 5:

a) There are cars which are mine.

b) Among them there is only one which is small.

c) This same car (the only one of mine) was imported from Japan.

This interpretation may be formulated in terms of set theory as follows:

$$I(5) : a_3 \supset n \cap a_1 \subset a_2$$

Model 5 represents the derivation tree for this interpretation.

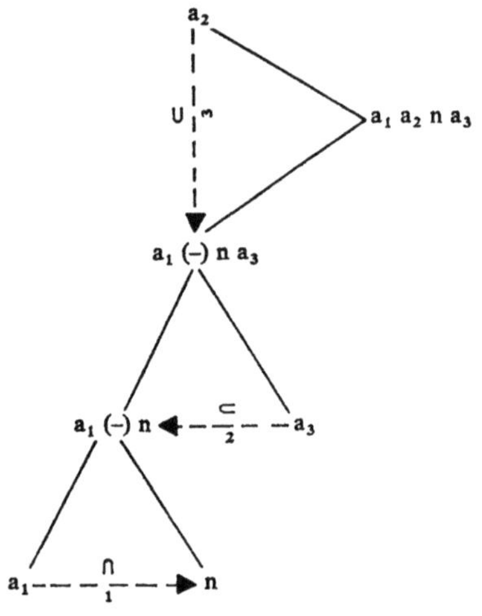

Model 5

In model 6, which also corresponds to interpretation 5, we have divided the constituent /a_1 a_2 a_3/ into two two-member immediate constituents, as it is not necessary to bring out a hierarchical difference between the two properties a_2 and a_3, since according to 5 either a_2 or a_3 is sufficient to identify *n*.

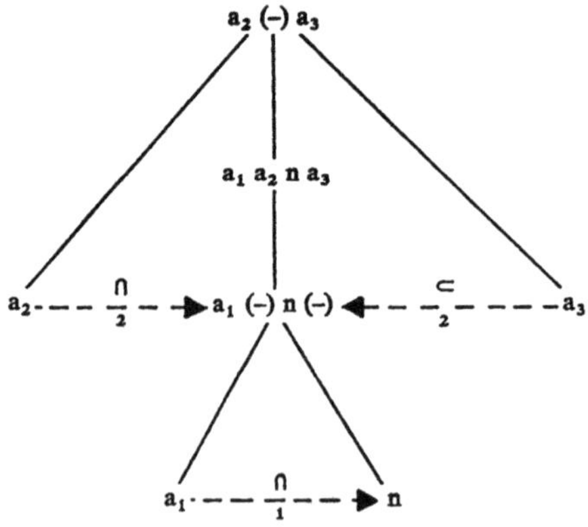

Model 6

The fact that our example can be interpreted semantically in several different ways should be taken into account when it is translated, either by a human translator or by a machine, into other languages where these ambiguities are not possible. To clarify any ambiguity considered to be due to different transformational processes undergone by different entry constructions, we can make use of transformational analysis. Here we shall apply transformational theory to interpretations 3 and 5.

$$\begin{aligned}
\mathrm{I}(3)\!: \; & T^{op}_{\cap a_1}: n \cap a_2 \Rightarrow n \cap a_2 \cap a_1 \\
& T^{op}_{\subset a_3}: n \cap a_2 \cap a_1 \Rightarrow n \cap a_2 \cap a_1 \subset a_3 \\
\mathrm{I}(5)\!: \; & T^{op}_{\supset a_3}: n \cap a_1 \Rightarrow a_2 \supset n \cap a_1 \\
& T^{op}_{\subset a_3}: a_2 \supset n \cap a_1 \Rightarrow a_2 \supset n \cap a_1 \subset a_3
\end{aligned}$$

Example 12: In this example we shall try to demonstrate a possible method of procedure.

a) *First Postulate:* All the words which fit the blank of pattern (10) belong to set *A*.

"John speaks English ________" (10)

b) *Proof of this Postulate:* Find if there are any words which fit the blank of pattern (10). There is, for example 'rapidly'.

c) *Second Postulate:* All the words which fit the blank of pattern (11) belong to set *B*.

"John is a ________ speaker" (11)

d) *Proof:* See if there are any words which fit pattern (11). There is, for example 'good'.

e) *Third Postulate:* All the words which fit both pattern (10) and pattern (11) belong to set *C*.

f) *Proof:* See if there are any words which fulfil this condition. There is, for example 'fast'.

g) *Conclusion:* Therefore there is an intersection between sets *A* and *B*.

h) *Fourth Postulate:* All the words which fit the blank of pattern (10) but not the blank of pattern (11), belong to set *D*.

i) *Proof:* See if there are any words which satisfy this condition. There are some, for example all the words in '-ly' derived from adjectives, 'perfectly' etc. 'Perfectly' does not fit pattern (11).

j) *Fifth Postulate:* All the words which fit the blank of pattern (11) but not the blank of pattern (10) belong to set *E*.

k) *Proof:* See if there are any words which fulfil this condition. There is, for example 'excellent'.

We have thus obtained five different sets, among which there is one intersection and two differences. We may illustrate this in the following diagram, with the corresponding formulae:

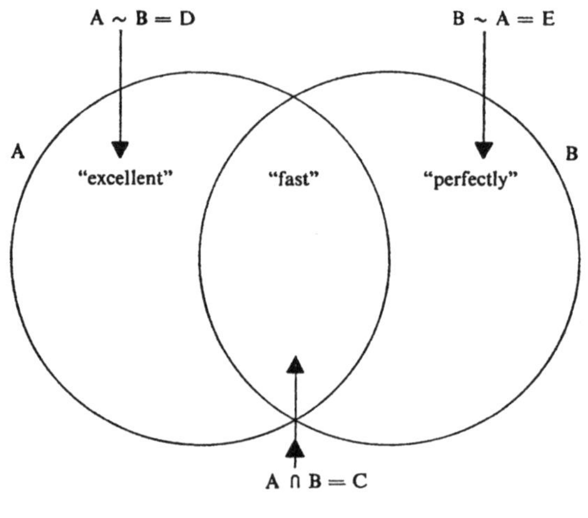

$A = \{a \mid a \text{ fits into (10)}\}$
$B = \{b \mid b \text{ fits into (11)}\}$
$C = \{c \mid c \text{ fits into (10) and (11)}\}$
$D = \{d \mid d \text{ fits into (10) only}\}$
$E = \{e \mid e \text{ fits into (11) only}\}$

Example 13: In Spanish the phonemes /P/ and /B/ have a set of definite phonetic features.[7]

The phoneme /P/ gives us the following set:

/P/ = {non-vocalic, consonantal, diffuse, grave, oral, interrupted, unvoiced}

The phoneme /B/ gives us the following set:

/B/ = {non-vocalic, consonantal, diffuse, grave, oral, voiced}

The common base of these two phonemes is the intersection:

/P/ ∩ /B/ = {non-vocalic, consonantal, diffuse, grave, oral}

The different characteristics of each of the phonemes can be represented by the following differences: The set of the difference:

/P/ ~ /B/ = {interrupted, unvoiced}

includes the distinctive features of the phoneme /P/, and the set of the difference:

/B/ ~ /P/ = {non-vocalic, voiced}

includes the distinctive features of the phoneme /B/.

We can see this in the following diagram:

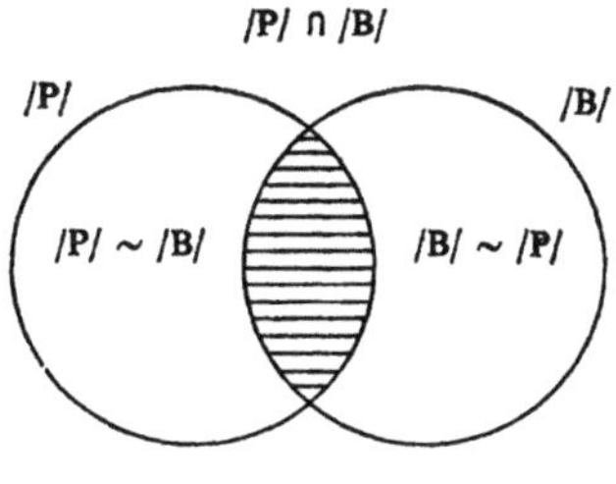

We can see from the diagram that the intersection or common base of the two phonemes is a subset of the union of the two sets /P/ and /B/ as well as a subset of each of the two sets.

[7] See: E. Alarcos, *Fonologia Española* (Madrid, 1961), p. 173, and R. Jakobson and M. Halle, *Fundamentals of Language* (The Hague, 1956).

e. *Union of a Finite Family of Sets*

Let us take sets A_1, A_2, ... A_n; we call the union of such sets, and write it $\bigcup_{i=1}^{n} A_i$, the set of elements which belong to at least one of these sets. That is:

$$x \in \bigcup_{i=1}^{n} A_i = A_1 \cup A_2 \cup \ldots \cup A_n \text{ means that } x \in A_1 \text{ or}$$
$$x \in A_2 \text{ or } \ldots x \in A_n$$

Example 14: In German, all nouns make up a set formed by the union of set A_1 of all the masculine nouns, set A_2 of all the feminine nouns, and set A_3 of all the neuter nouns. We have:

$$A_1 \cup A_2 \cup A_3 = \bigcup_{i=1}^{3} A_i$$

f. *Intersection of a Finite Family of Sets*

Given the sets A_1, A_2, ... A_n, the intersection of these sets, which we write $\bigcap_{i=1}^{n} A_i$ is the set of elements which belong simultaneously to all the A_i. That is to say:

$$\text{If } x \in \bigcap_{i=1}^{n} A_i = A_1 \cap A_2 \cap \ldots \cap A_n, \text{ then } x \in A_1, x \in A_2,$$
$$\ldots x \in A_n$$

Example 15: Let us take the following sets of verbs:

a) Verbs which only fit a structure represented by pattern (12):

"Noun + ________." (12)

"John died."

We shall denote these verbs v_1. They are verbs which we may call *necessarily monovalent* since they cannot take an object.

b) Verbs v_2 which only fit a structure represented by pattern (13).

"Noun + ——— + direct object." (13)
"John hates the teacher."

We may call these verbs *necessarily bivalent.*

c) Verbs v_3 which only fit a structure represented by pattern (14).

"Noun + ——— + direct object + indirect object."
"John gives the book to the teacher."

The order of words is not fixed. We can also include sentences like:

"John gives me the book."
"John gives the teacher the book."

We may call these verbs *necessarily trivalent.*

d) Verbs v_4 which fit both structures represented by patterns (12) and (14), i.e.:

"John is eating."
"John is eating an apple."

We may call these verbs *optionally bivalent.*

e) Verbs v_5 which fit the structures represented by patterns (13) and (14), i.e.:

"John bought a book."
"John bought me a book."

We may call these verbs *optionally bi-trivalent.*

f) Verbs v_6 which fit the structures represented by patterns (12), (13) and (14), i.e.:

"John is reading."
"John is reading a letter."
"John is reading me a letter."

We may call these verbs *optionally multivalent.*

g) Verbs v_7 which fit the structure represented by patterns (14) and (12), but not that of pattern (13). It appears that we cannot find any verb in English which fulfils these conditions, and thus this set is empty, that is to say a category without elements (see definition of an empty set, page 4).

Thus we have a family of three sets in intersection, which gives us a total of seven sets: four intersections and three differences. This can be illustrated by means of the following diagram:

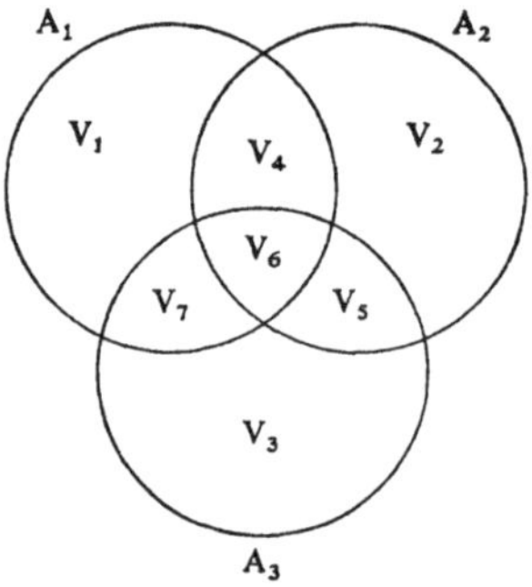

$$A_1 \cap A_2 \cap A_3 = \bigcap_{i=1}^{3}$$

$V_1 = \{v_1 \mid v_1$ is necessarily monovalent$\}$
$V_2 = \{v_2 \mid v_2$ is necessarily bivalent$\}$
$V_3 = \{v_3 \mid v_3$ is necessarily trivalent$\}$
$V_4 = \{v_4 \mid v_4$ is optionally bivalent$\}$
$V_5 = \{v_5 \mid v_5$ is optionally bi-trivalent$\}$
$V_6 = \{v_6 \mid v_6$ is optionally multivalent$\}$
$V_7 = \emptyset$ (empty set)

Example 16: In verbal communication between two individuals we must distinguish between the active inventory of symbols, typical constructions (structural signs), that is to say those symbols which each individual disposes of to form signals (which transmit messages), and the passive inventory of symbols, that is to say those symbols which the individual who is receiving a signal recognises and uses to decode it, but which he does not use to send a signal. Thus each individual has his own language, called an *ideolect.* We can see graphically in the following diagram the communication between two individuals. The small circles represent the active inventories of each one, the larger circles the passive inventories.

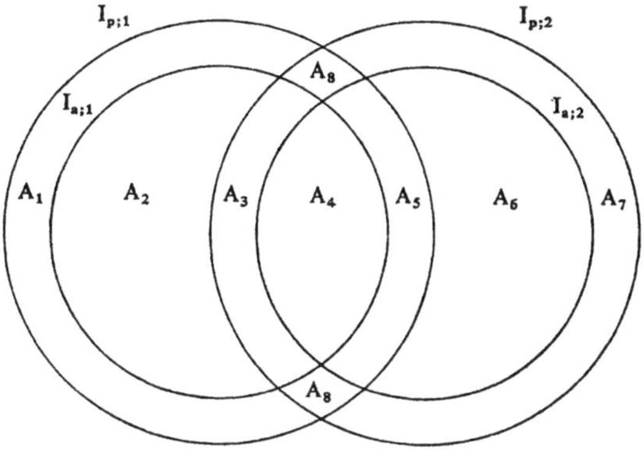

Active inventories: $I_{a;1}, I_{a;2}$

Passive inventories: $I_{p;1}, I_{p;2}$

$$A_1 = \{a_1 \mid a_1 \text{ is an element of } I_{p;1} \text{ only}\}$$
$$A_2 = \{a_2 \mid a_2 \text{ is an element only of } I_{p;1} \text{ and } I_{a;1}\}$$
$$A_3 = \{a_3 \mid a_3 \text{ is an element only of } I_{p;1}, I_{a;1} \text{ and } I_{p;2}\}$$
$$A_4 = \{a_4 \mid a_4 \text{ is an element of } I_{p;1}, I_{a;1}, I_{p;2}, I_{a;2}\}$$
$$A_5 = \{a_5 \mid a_5 \text{ is an element only of } I_{p;1}, I_{p;2} \text{ and } I_{a;2}\}$$
$$A_6 = \{a_6 \mid a_6 \text{ is an element only of } I_{p;2} \text{ and } I_{a;2}\}$$
$$A_7 = \{a_7 \mid a_7 \text{ is an element only of } I_{p;2}\}$$
$$A_8 = \{a_8 \mid a_8 \text{ is an element only of } I_{p;1} \text{ and } I_{p;2}\}$$

This case may be illustrated by the following situation: An Italian who speaks only his native language is speaking to a person who speaks only Spanish. In the process of communication the following cases may arise:

1) The Italian uses an Italian word which, being similar in form and meaning to a Spanish word, is understood by the Spanish-speaking person. This word would be located in set A_3.

2) The Italian uses an Italian word whose form and meaning is identical in Spanish. It will be located in set A_4.

The reader may formulate the other cases.

Example 17: Note the restrictive effect of the attribute in the expression: "John's friend", as compared to the descriptive effect of the attribute in the expression: "dear friend". The first expression — denoting the nucleus as *n* and the attribute as *a* — may be formulated as follows:

$n \cap a$

The second expression as follows:

$a \supset n$ (Among the people I like there will also be my friend)

This is important for the placing of the adjective in Spanish. Note the difference between:

"el vivo Pérez": $a \supset n$
and "el Pérez vivo": $a \cap n$

g. *Associative Law*

A method of combining objects two at a time is *associative* if the result of the combination of three objects (order being preserved) does not depend on the way in which the objects are grouped. If the operation is denoted by and the result of combining *x* and *y* by $x \circ y$ then:

$$(x \circ y) \circ z = x \circ (y \circ z)$$

for any *x*, *y*, and *z* for which the operation is defined.

In the same way this law applies to operations of union and intersection of sets.

$$(A \cup B) \cup C = A \cup (B \cup C)$$
$$(A \cap B) \cap C = A \cap (B \cap C)$$

Example 18: Let *A* be the set of the English words *a* which fit the blank of pattern (15), *B* the set of words *b* which fit the blank of pattern (16), and *C* the set of words *c* which fit the blank of pattern (17):

"The ________ was very interesting." (15)
"We ________ every day." (16)
"He is a ________ actor." (17)

Any word which fits all three patterns is of course an element of the intersection of the three sets, for example the word 'play', which can have the function of a Class I word, a Class II word, or a Class III word.[8]

> "The play was very interesting."
> "We play every day."
> "He is a play-actor."

Elements exist which can complete patterns (15) and (16), but not all of them can also complete pattern (17). This means first making up the set of the intersection of sets *A* and *B* and then making up the intersection of this intersection with set *C*. The word 'play' will be found in this final intersection.

But elements also exist which can complete patterns (16) and (17), but not all these can also complete pattern (15). This means first making up the intersection of sets *B* and *C*, and then making the intersection of this intersection with set *A*. In this final intersection our word 'play' appears again. This may be illustrated by means of the following diagram:

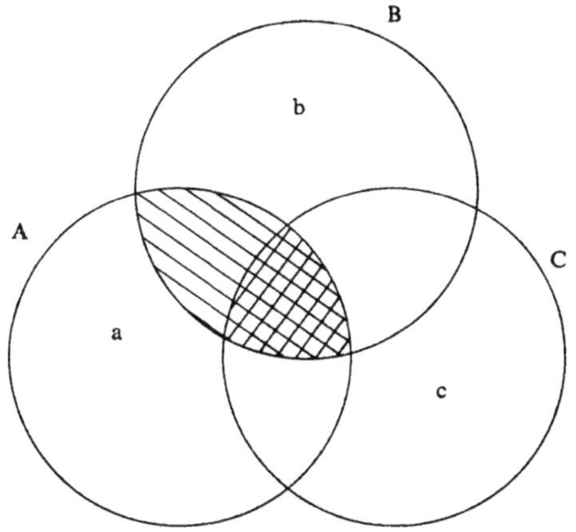

"play" $\in (A \cap B) \cap C$

[8] In the terminology of Charles Fries. To be more precise, as a Class III word 'play' belongs to a subset of this class, since its behaviour is not the same as that of every other Class III word; for example, it cannot be used a a predicative adjective.

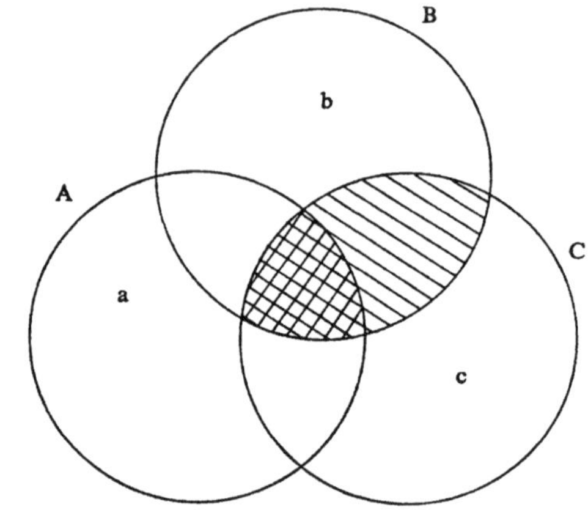

"play" $\in A \cap (B \cap C)$

h. *Distributive Law*

An operation is distributive relative to a rule of combination if performing the operation upon the combination of a set of quantities is equivalent to performing the operation upon each member of the set and then combining the results by the same rule of combination; for example, in arithmetic and algebra, the law which states that:

$$a(b + c) = ab + ac$$

The same law applies to set operations. We get two laws:

1) $A \cup (B \cap C) = (A \cup B) \cap (A \cup C)$
2) $A \cap (B \cup C) = (A \cap B) \cup (A \cap C)$

Example 19: Let *A* be the set of the lexical elements which fit pattern (18) and/or pattern (19), *B* the set of all adjectives and *C* the set of those adjective which fit pattern (20).

"John is a ________ runner." (18)
"John runs ________." (19)
"John is 20 years ________." (20)

The word 'fast' may be inserted in either of the two patterns (18) or (19). This word is thus an element of set *A*, which forms an intersection with set B. The words which may be inserted in pattern (20) form a subset of set B. One element of this subset is the adjective 'old'.

This may be seen in the following diagrams:

$A \cup (B \cap C)$ is the totally shaded area

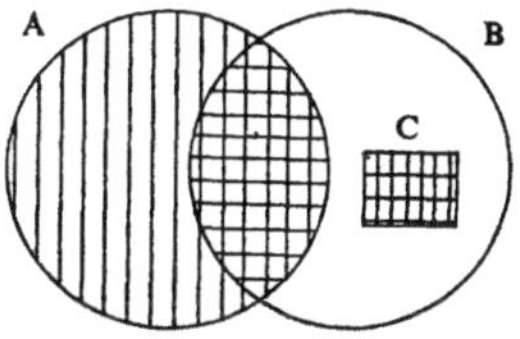

$(A \cup B) \cap (A \cup C)$ is only the area shaded with horizontal *and* vertical lines

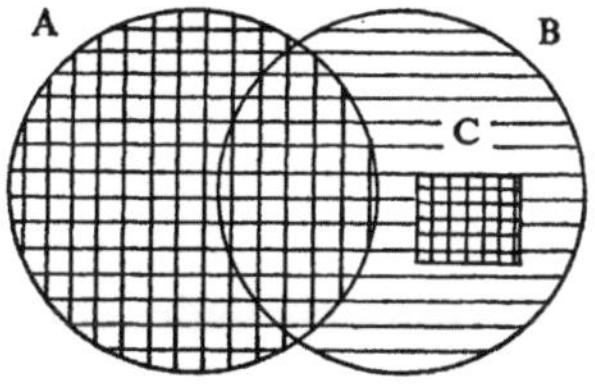

Example 20: Let A be the set of all the Class IV words (adverbs), B the set of all the Class I words (nouns), and C the set of all the Class III words (adjectives). Any element which fits patterns (18) and (19), for example the word 'fast', is an element of the set $A \cap (B \cup C)$, and is found in the area with vertical and horizontel shading in the first diagram and in the area with vertical shading in the second:

$A \cap (B \cup C)$

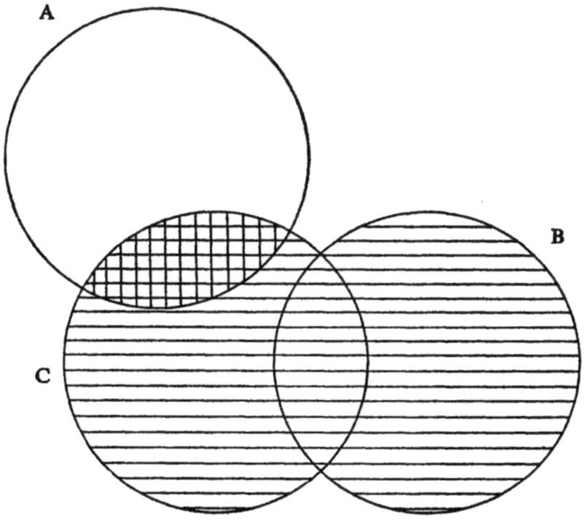

(A ∩ B) ∪ (A ∩ C)

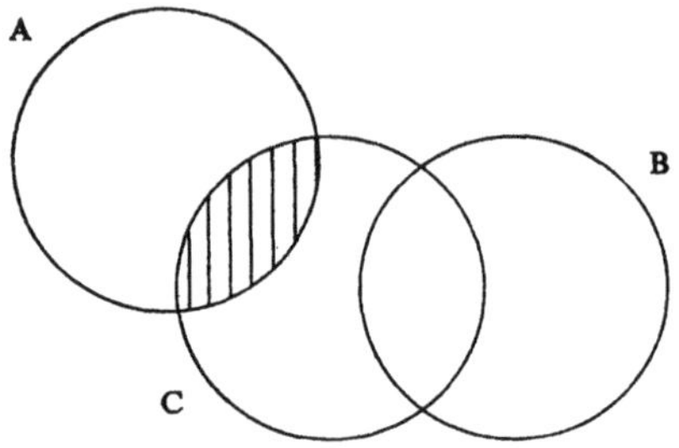

In the example under consideration the intersection of sets *B* and *A* is an empty set. However, if we make the condition that the elements which fit pattern (21) *and* also pattern (22) belong to a certain set, the two sets *B* and *A* will make an intersection and the diagram would have to be changed:

"John sings ________." (21)
"There is a ________ here." (22)

The word 'well' fits both patterns.

6. PROPERTIES OF BINARY RELATIONS

a. *Ordered Pairs*

An ordered pair consists of two elements, for example *a* and *b*, of which one is designated as the first element, and the other as the second. This is denoted in parenthesis: (a,b)

Example 21: In Spanish, the pattern 'auxiliary plus main verb' has a fixed order, as in sentence (23); the order which appears in (24) is not syntactically possible:

"Juan ha comprado un libro." (23)
"Juan comprado ha un libro." (24)

b. *Cartesian Product*

Given the sets *A* and *B* we call the *Cartesian product* of them the set of all the pairs (x,y) where $x \in A$, $y \in B$. We denote this:

$$A \,.\, B = \{(x,y) \mid x \in A,\ y \in B\}$$

in which *x* is the first component and *y* the second component.

Example 22: Let *A* be the set of all the auxiliary verbs *x*, and *B* the set of all the main verbs *y* together with which the auxiliary verbs *x* can occur in the order 'auxiliary verb plus main verb', written $x \prec y$, then the total of the sentences which may be formed with 'auxiliary plus main verb' represents a Cartesian product:

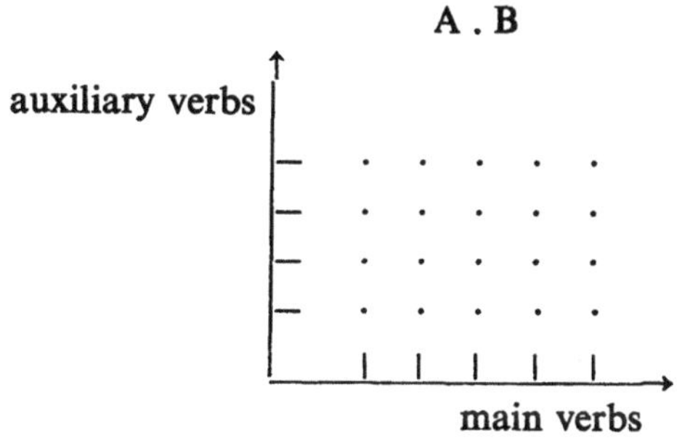

c. *Binary Relations*

Given the sets *A* and *B* and a property which the pairs $(x,y) \in A \,.\, B$ may or may not have, we say that this property defines a *binary relation* between the elements of those sets. We say that two elements $x \in A$, $y \in B$ are related if the pair (x,y) has this property. This is written:

$$(x,y) \in R$$

in which R expresses the relation.

We define the *domain* of the relation R as the set of the elements of *A* which are first components of the pairs which are in relation R.

We define the *range* of the relation R as the set of the elements of B which are second components of the pairs which are in the relation R.

d. *Classification of Binary Relations*

1) *Inverse Relation:* Given two sets *A* and *B* which are in the relation R, we call the *inverse* relation of R, written R^{-1}, the set defined as follows:

$$R^{-1} = \{(y,x) \mid (x,y) \in R\},$$

Example 23: Let *A* be the set of auxiliary verbs *a*, and *B* the set of the main verbs *b* with which the auxiliary verbs *a* combine, we have the following relation:

$$R = \{(a,b)\ a \text{ is an auxiliary of } b\},$$
$$R^{-1} = \{(b,a)\ b \text{ is a main verb that goes with } a\}$$

Other relations and their inverse forms are as follows:

a) "x is the subject of y"	"y is the predicate of x"
b) "x is governed by y"	"y governs x"
c) "x precedes y"	"y follows x"

as may be seen in the following Spanish expressions:

a) "Juan (x) duerme (y)."
b) "entre (y) tú (x) y yo (x)." (The preposition "entre" requires the nominative form of the pronoun, unlike other prepositions; "contigo" or "conmigo".)
c) "auxiliary verb + main verb."

2) *Reflexive Relation:* A binary relation is *reflexive* if for $\forall x \in A$, $(x,x) \in R$, or $x\,R\,x$.

Example 24: Relation "*a* acts on *a*" in the cases of reflexive actions in Spanish:

"Juan (a) se (a) peina."

It should be noted that the English translation of this sentence is not reflexive:

"John combs his hair."

Also, not every expression defined grammatically as reflexive is reflexive in the sense of a reflexive relation as we conceive it here. Thus in Spanish:

"Juan se ríe."

does not contain a reflexive relation. In this sense 'reirse' is not a reflexive verb. It is essential to formulate the relation clearly.

3) *Irreflexive Relation:* A binary relation R is *irreflexive* in set *A* if, for every *x* in *A*, x R x is not true. We write:

$$\text{R is irreflexive in A} \leftrightarrow (\forall x \in A \rightarrow -(x\ R\ x))$$

Example 25: An irreflexive relation is as follows: "*a* never acts on *a*". This relation is established with verbs which fulfil this condition, as in the following example:

"John drinks."

4) *Non-reflexive Relation:* A binary relation R is *non-reflexive* in set *A* if there is at least one *x* for which x R x is not true.

Example 26: An example of a non-reflexive relation is as follows: "*a* can also act on *b*". We can establish this relation with verbs which fulfil these conditions, for example the following:

"Juan (a) estima a su amigo (b)."
"Juan (a) se (a) estima (a sí mismo)."

5) *Symmetric Relation:* A binary relation is *symmetric* if

$$(x,y) \in R \rightarrow (y,x) \in R.$$

Example 27: Every sentence conceived as such must have, explicitly or implicitly, a subject and a predicate. The relation is: "the subject must be complemented by a predicate, and the predicate by a subject"; one justifies the existence of the other (principle of polarity).

Let set *A* contain as elements all the words which can fulfil the syntactical function of the subject (w_s) and all the words which can fulfil the syntactical function of the predicate (w_p). The relation is one of polarity (P). We write:

$$\text{P is symmetric in A} \leftrightarrow \forall\, w_s \in A \ \&\ \forall\, w_p \in A \ \&\ w_s\, P\, w_p \rightarrow w_p\, P\, w_s$$

6) *Asymmetric Relation:* A relation R is *asymmetric* in set *A* if for each *x* and *y* in *A*, when x R y, y R x is not true. We write:

$$\text{R is asymmetric in A} \leftrightarrow \forall\, x \in A \ \&\ y \in A \ \&\ x\, R\, y \rightarrow -(y\, R\, x)$$

Example 28: An auxiliary verb is auxiliary with reference to a main verb. Let set *P* contain as elements the auxiliary verbs (x) and the verbs which can only be main verbs (r). The relation between *x* and *r* is that "*x* syntactically helps *r*" (and not the other way round).

7) *Anti-symmetric Relation:* A relation R is *anti-symmetric* in set *A* if for each *x* and *y* in *A*, when x R y and y R x, x = y.

Example 29: In any sentence in Spanish, a grammatical person (noun, pronoun, or one of these implicitly) must correspond to a given finite verb-ending; and a specific finite verb-ending corresponds to each pronoun or noun. Thus the grammatical person is represented in a finite verb-ending and also in the corresponding noun or pronoun.

Let set *Y* contain as elements the meanings of person, *u*, of the nouns and pronouns, and the meanings of persons expressed in the verb-endings, *v*. The relation Z between *u* and *v* is that "*u* corresponds to *v* as regards person, and *v* corresponds to *u* as regards person in such a way that both express the same person. For example in Spanish:

"Los amigos	(3rd person plural)
vienen."	(3rd person plural)

We write:

Z is anti-symmetric in Y $\leftrightarrow \forall$ u $\in$ Y & v $\in$ Y & u Z v & v Z u $\rightarrow$ u = v

8) *Transitive Relation:* A relation R is *transitive* in set *A* if, for each *x*, *y* and *z* in *A*, provided that x R y and y R z, then x R z.

Example 30: A relation which produces the substantivation of the adjective in Spanish.

Let *N* be the set of the articles *d*, the nouns *n*, and the adjectives (not used as epithets) *a*. The relation is one of antecedence (A). We write:

A is transitive in N $\leftrightarrow \forall$ d $\in$ N & n $\in$ N & a $\in$ N & d A n & n A a $\rightarrow$ d A a

For example:

The article precedes the noun: "el carro"
The noun precedes the adjective: "carro grande"
The article can precede the adjective: "el grande"

9) *Intransitive Relation:* A relation R is intransitive in set *A* if for each *x*, *y* and *z* in *A*, when z R y and y R z, x R z is not true.

Example 31: The relation which in German requires that in a subordinate clause introduced by a conjunction the verb can never precede the subject:

"conjunction + subject + verb"
"dasz er käme"

Let *S* be the set of the subordinate conjunctions *c*, the nouns or personal pronouns *p* and the verbs *v*. The relation is one of antecedence. We write:

A is intransitive in S $\leftrightarrow \forall$ c $\in$ S & p $\in$ S & v $\in$ S & c A p & c A p & p A v $\rightarrow$ — (c A v)

The construction "..., dasz käme er" would be ungrammatical.

10) *Equivalence Relation:* A relation R in set A is called an *equivalence relation* if

a) R is reflexive,
b) R is symmetric, and
c) R is transitive.

Example 32: Let us suppose that three people studying English have to pronounce the sentence, "I hit it". In this sentence the vowel in the verb is short. If it is lengthened, the meaning of the sentence is changed to "I heat it". For a fourth person to hear and interpret the sentence with the first meaning, the three people must pronounce a short vowel, allowing for slight variation in length. Each time they repeat the sentence, the length of the vowel must not exceed certain limits. In terms of sets:

If A is the set of the phonemes /I/ (= short [i]), then each phoneme /I/ produced by the three people must always be the same (condition of reflexive relation); if A is the set of the phonemes /I/ produced by person 1, and B is the set of the phonemes /I/ produced by person 2, the phonemes produced by person 1 must be the same as those produced by person 2, and vice versa, since otherwise the meaning is changed (condition of symmetry); and if C is the set of the phonemes /I/ produced by person 3, then if the phonemes /I/ produced by person 1 are the same as those produced by person 2, and the phonemes /I/ produced by person 2 are the same as those produced by person 3 (as they must be to avoid changing the meaning of the sentence), then the phonemes /I/ produced by person 1 are the same as those produced by person 3 (condition of the relation of transitivity).

II

FUNCTIONS

1. GENERAL DEFINITION

Let *A* and *B* be two non-empty sets; we say that *f* is a *function* or *mapping* of *A* into *B*, and we write it: A → B. For each x ∈ A there is only one element y ∈ B, such that *y* is assigned to *x* by the function *f*:

$$f: A \rightarrow B \leftrightarrow \forall x \in A\ \exists y \in B \mid y = f(x)$$

The symbol ∃ means 'exists'. Set *A* is called the *domain* of function f, and set *B*, the *range* of function f. We can illustrate this by means of the following diagram:

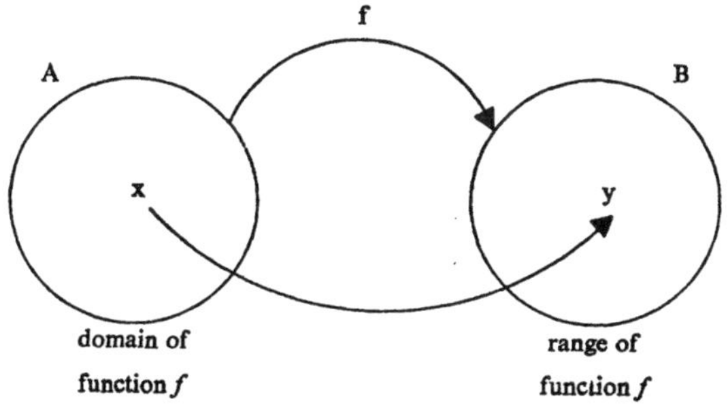

Example 33: Let *N* be the set of all the Spanish nouns *n*, arranged as masculine singular nouns ($n_{m;s;1}$, $n_{m;s;2}$, ... $n_{m;s;n}$) masculine plural nouns ($n_{m;p;1}$, $n_{m;p;2}$, ... $n_{m;p;n}$), feminine singular nouns ($n_{f;s;1}$, $n_{f;s;2}$, ... $n_{f;s;n}$), and feminine plural nouns ($n_{f;p;1}$, $n_{f;p;2}$, ... $n_{f;p;n}$); let *A* be the set of all the Spanish definite articles which do not form a contraction with a preposition ($a_{m;s}$, $a_{m;p}$, $a_{f;s}$, $a_{f;p}$). *A*

is a function of *N* if for each $n \in N$ there is an element $a \in A$ such that *a* is assigned to *n* by the function of *agreement* (c). We write:

$$c: N \rightarrow A \leftrightarrow \forall n \in N, \exists a \in A \mid a = c(n)$$

$a = c\,(n_{g;u})$ (*a* is a function of agreement determined by the gender (g) and the number (u) of the noun (n).)

2. TYPES OF FUNCTIONS

a. *Onto-mapping*

A function $f: A \rightarrow B$ is said to be an *onto-mapping* if for every $y \in B$ there is a $x \in A$ such that $y = f(x)$. In other words, set *B* does not contain elements which are *not* images of the elements of set *A*, according to some function.

Example 34: Agreement between nouns and articles. (Cf. example 33.)

b. *Into-mapping*

This is the case when set B also contains elements which are *not* images of the elements of set *A*, according to some function. We write:

$$f(A) \subset B$$

Example 35: Let *N* be the set of Spanish singular nouns (n_s), and all Spanish plural nouns (n_p), and *P* the set of all the Spanish finite verb-endings. *P* is a function of *N* if for each $n \in N$ there is an element (verb-ending) which belongs to set *P* such that this element is assigned to *n* by the function of agreement *c*, and such that there are also in set *P* verb-endings which do not agree with any element *n* of *N*.

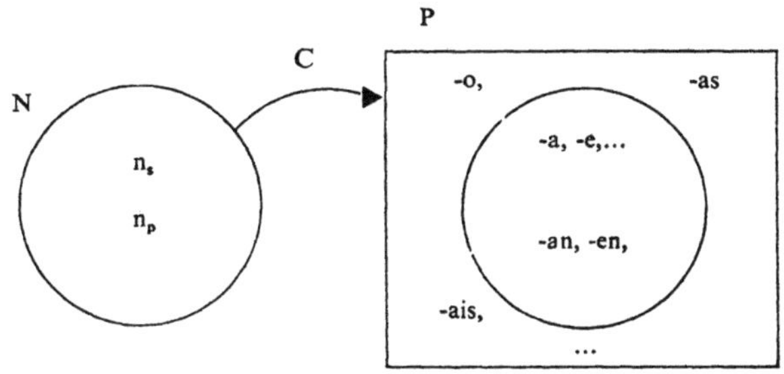

$c(N) \subset P$ = all endings that correspond to the 3rd person singular and to 3rd person plural

c. *Biunique Function*

A function is called *biunique* if for $x_1 \neq x_2$ we have $f(x_1) \neq f(x_2)$.

Example 36: Function of agreement (see example 33).

d. *Non-Biunique Function*

If $x_1, x_2 \in A \mid x_1 \neq x_2 \rightarrow -(f(x_1) \neq f(x_2))$, then f is *not* biunique.

Example 37: Let N be the set of the singular forms of the Spanish masculine nouns which end in *-o* ($n_{m;-o;s}$: e.g. 'pueblo'), of the plural forms of the Spanish masculine nouns which end in *-o* ($n_{m;-o;p}$), masculine nouns which end in *-a* in the singular ($n_{m;-a;s}$), and the same in the plural ($n_{m;-a;p}$), feminine nouns which end in *-a* in the singular ($n_{f;-a;s}$), the same in the plural ($n_{f;-a;p}$), the feminine nouns which end in *-o* in the singular ($n_{f;-o;s}$), and the same in the plural ($n_{f;-o;p}$); and A the set of the definite articles, then the function between N and A is not biunique, and we write:

$$\text{If } n_{m;-o;s}, n_{m;-o;p}, \ldots n_{f;-o;p} \in N \mid n_{m;-o;s} \neq n_{m;-o;p} \neq \ldots n_{f;-o;p} \rightarrow -(c(n_{m;-o;s}) \neq c(n_{m;-a;s}), c(n_{f;-a;s}) \neq c(n_{f;-o;s}), c(n_{m;-o;p}) \neq c(n_{m;-a;p}), c(n_{f;-a;p}) \neq c(n_{f;-o;p}))$$

Examples: "*el* sombrer*o*, *el* sof*á*, *la* mes*a*, *la* man*o*; *los* sombrer*os*, *los* sof*ás*, *las* mes*as*, *las* man*os*"

e. *Inverse Function*

If a function f: A → B is an onto function and biunique we can determine another function f^{-1}: B → A, called the *inverse function*:

$$f^{-1}(B) = x \in A \mid y = f(x)$$

Example 38: In Spanish we have the following phonemic contrasts:

/K/ : /G/
/P/ : /B/
/T/ : /D/

The distinctive feature of this correlation is voicing. We can then formulate the following function: By means of devoicing we make each voiced phoneme correspond to an unvoiced one.

Let *A* be the set of the unvoiced occlusive consonants /K/, /P/, /T/, commonly called *x*, and *B* the set of the voiced occlusive consonants /G/, /B/, /D/, commonly called *y*, then set B is linked to set *A* by means of the function *v* of voicing, that is to say:

$$\text{Function } v: A \rightarrow B \leftrightarrow \forall x \in A \;\; \exists y \in B \mid y = v(x)$$

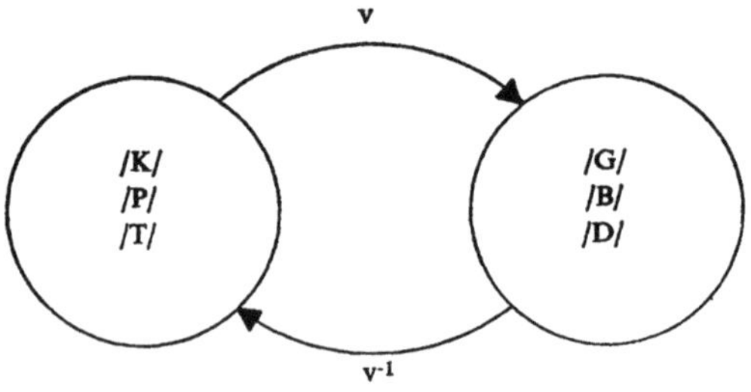

As we can see, an onto-mapping is involved.

In addition, since for $x_1 \neq x_2$ we have $v(x_1) \neq v(x_2)$, it is a biunique function. And since it is a biunique onto-function we can define the inverse function:

$$v^{-1}(B) = x \in A \mid y = v(x)$$

In this case, the inverse function would be *devoicing*.

Example 39: Function of agreement as shown in example 33:

$$a = c(n_{g;u}) \qquad c^{-1}(A) = \{n \in N \mid a = c(n)\}$$
$$n_{m;s} = c^{-1}(\text{"el"})$$

For example the sentence in Spanish "El ________ está allí" must be completed by a masculine singular noun.

f. *Equal Functions*

If *f* and *g* are functions defined by the same domain D and if $f(a) = g(a)$ for every $a \in D$, then the functions *f* and *g* are *equal.*

Example 40: Let *N* be the set of the Spanish nouns $n_{m;s}$, $n_{n;p}$, $n_{f;s}$, $n_{f;p}$, and *A* the set of the Spanish definite articles and the Spanish adjective-endings in a vowel. The noun requires that in any sentence with an article and an adjective, these should agree with it in gender and number. The function is one of agreement. Since there are two agreements of the same type (both refer to gender and number), the two functions are equal.

$$d(= \text{form of the article}) = c_1(n_{g;u})$$
$$a(= \text{adjective-ending}) = c_2(n_{g;u})$$

g. *Constant Function*

A function *f* of A into B is called a *constant* function if the same element $b \in B$ is assigned to every element in A:

$$f: A \rightarrow B$$
$$f(a) = \text{constant } \forall a \in A$$

Example 41: In English, all nouns ending in a voiceless consonant (except those in /-s/, /-sh/ etc.) take a voiceless form of the /S/ phoneme in the plural: /hæts/. The same holds for verbs in the third person singular of the present tense.

Let *A* be the set of the English nouns ending in a voiceless consonant (except for those in /-s/ etc.), and the English verbs ending in a voiceless consonant in the present tense (except those in /-s/ etc.), and let *B* be the set of only one element; the allophone [s] of the /S/ phoneme. The function is constant:

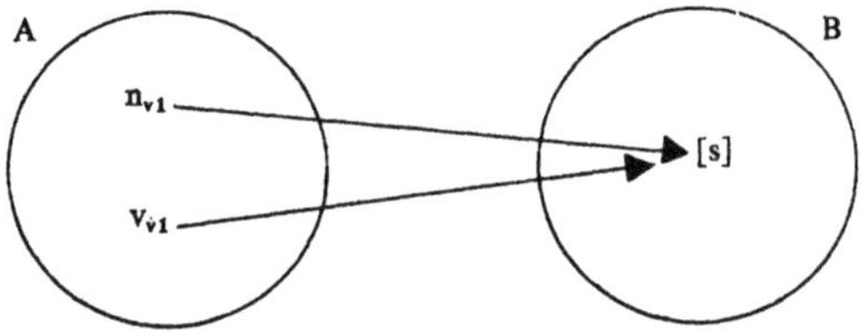

h. *Product Function*

If f: A → B and g: B → C then we define a function (g . f): A — C as (g . f) (a) ≡ g(f(a)). The symbol ≡ means 'equal by definition'.

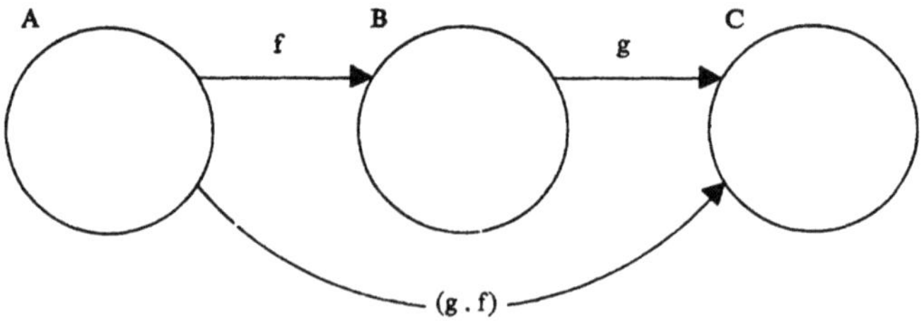

Example 42: In classical Sanskrit the occlusive consonants form both a correlation of aspiration and a correlation of voicing, giving four-term bundles.

$$\begin{array}{ccc} K - K^h & P - P^h & T - T^h \\ | \quad\quad | & | \quad\quad | & | \quad\quad | \\ G - G^h & P - P^h & T - T^h \end{array}$$

Let *A* be the set of the voiceless occlusives /K/, /P/, /T/, together denominated *u*, *B* the set of the voiced occlusives /G/, /B/, /D/, together denominated *z*, *C* the set of the aspirate voiced occlusives /Gʰ/, /Bʰ/, /Dʰ/, together denominated *y*, *D* the set of aspirate voiceless stops /Kʰ/, /Pʰ/, /Tʰ/, together denominated *x*. Set *C* is linked to set *A* via set *B*, by means of the functions of aspiration (a) and voicing (v), that is to say we have the following product function:

a) $z = v(u)$
$y = a(z)$
$y = a(v(u)) \equiv (a \, . \, v)(u)$ (Cf. the following diagram.)

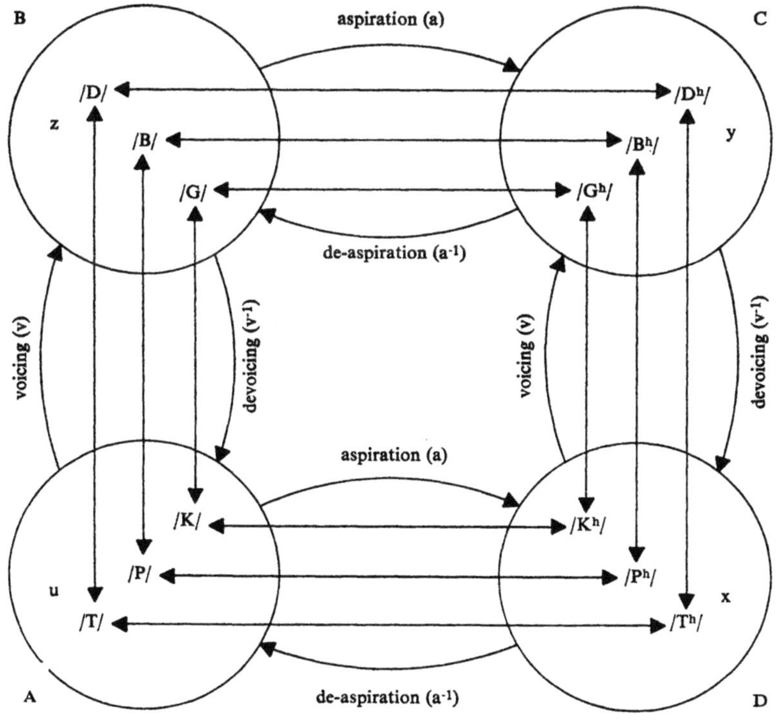

As we are dealing with a biunique onto-function we can define the inverse function as follows:

$$v^{-1}(B) = u \in A \mid z = v(u)$$
$$a^{-1}(C) = z \in B \mid y = a(z)$$
$$\text{I. } u = v^{-1}(z)$$
$$\text{II. } z = a^{-1}(y)$$

By substitution of II in I we get:

$u = v^{-1}(a^{-1}(y)) \equiv (v^{-1} . a^{-1})(y)$, which is the inverse function of the product function $y = (a . v)(u)$. The functions a^{-1} and v^{-1} are those of de-aspiration and devoicing. (Cf. diagram.)

We can further define the following functions:

b) $x = a(u)$
$y = v(x)$
$y = (v . a)(u)$, and its inverse function $u = (v^{-1} . a^{-1})(y)$

c) $y = v(x)$
$z = a^{-1}(y)$
$z = (a^{-1} . v)(x)$, and its inverse function $x = (v^{-1} . a)(z)$

d) $u = a^{-1}(x)$
$z = v(u)$
$z = (v . a^{-1})(x)$, and its inverse function $x = (a . v^{-1})(z)$

Example 43: In Japanese the negative of a verb is formed by means of the suffixes '-nai' or '-masen', and the past tense of these forms by means of the suffixes '-katta' or '-deshita', which are put after the previous suffixes:

"tabe-nai" (do not eat)	"tabe-masen"
"tabe-na-katta" (did not eat)	"tabe-masen-deshita"

The past tense negative is formed by double suffixation.

Let *A* be the set of the verbs *x* in their stem form which can take the suffix of negation, *B* the set of the same verbs *y* but with the suffix '-nai', and *C* the set of the same verbs *z* but with the suffixes '-na' and '-katta'; set *C* is linked to set *A* via set *B*, by means of the function of primary suffixation *p* and a secondary function, *s*, that is to say we have a product function:

$$y = p(x)$$
$$z = s(y)$$
$$z = s(p(x)) \equiv (s \,.\, p)(x)$$

i. *Inverse of a Function*

The function *f* ls not necessarily an onto-function:

$$f^{-1}(b) \text{ is the set } \{a \in A \mid f(a) = b\}$$

Example 44: Let *N* be the set of the Spanish nouns $n_{m;-o}$ and $n_{m;-a}$, and *A* the set of only one element, the definite article "el", the inverse of the function of concordance (c) is:

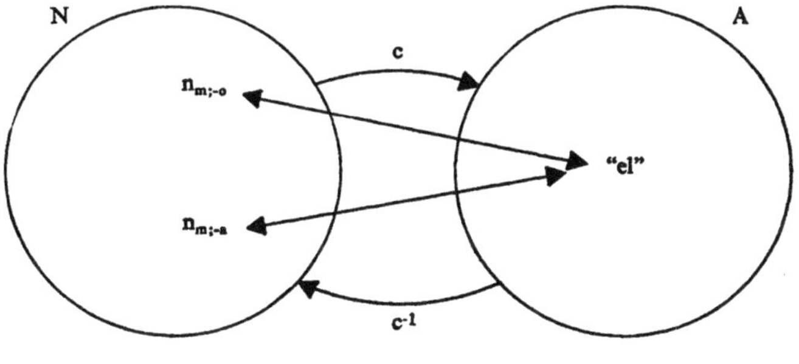

$$c^{-1}(el) = \{n \in A \mid c(n) = el\}$$

III

OTHER EXAMPLES OF THE APPLICATION OF SET THEORY TO LINGUISTICS

1. THE COMPARATIVE[1]

A syntactical construction is called *endocentric* when as a whole it belongs to the same formal class as one or more of its immediate constituents.[2] Thus the phrase "John the student" is endocentric in the sentence "John the student is ill" because the word "John" belongs to the same formal class as the phrase "John the student", that is to say it has the same structural function in the same sentence as the phrase "John the student"; "John is ill". A syntactical construction which represents a different formal class from that represented by any of its immediate constituents is called *exocentric*. Thus syntactically the construction "subject + predicate" is exocentric because both constituents together belong to the class of sentences.

In a construction "adjective + noun" we can consider the noun as the *nucleus* and the adjective as the *modifier*. It is an endocentric construction.

Example 45: Let us examine the endocentric construction of the Spanish sentence:

"Es un hombre más gordo que un cura."

We can write this sentence as follows:

$$n_h + \text{más} + s + \text{que} + n_s$$

[1] The following examples and their commentaries are based on H. Pilch's article "Comparative Constructions in English", *Language* 41.37 1965. This article is outstanding for its excellent approach to the linguistic side of the problem, and also for its adequate application of set theory.

[2] For theory of immediate constituents see footnote 5.

Let *N* be the set of all human beings *n* and *S* the set of all priests n_s. Therefore $S \subset N$. It is a transitive relation: *n* is more *s* than n_s. We write:

$$R = \{(n_h \in N \sim S, n_s \in S \subset N) \mid n_h \text{ es más s que } n_s\}$$

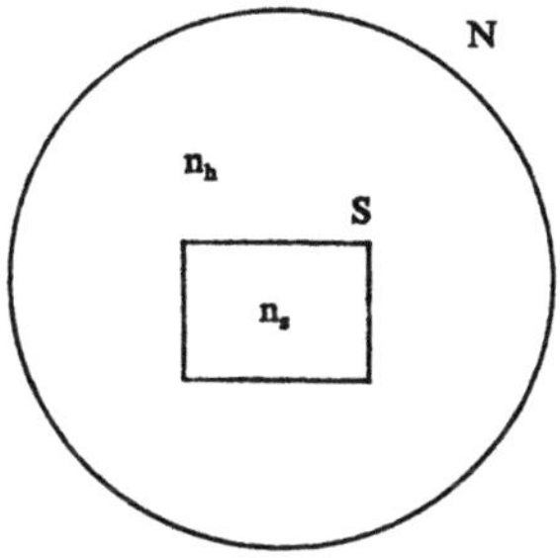

This formula shows the following restrictions:[3]

a) n_h is an indefinite element in the complement of subset *S*. This restriction prevents the formation of sentences like the following:

* "Es el hombre más gordo que un cura."

since here n_h is not an indefinite element.

b) n_s is a definite element in set *N*, in the sense that it forms part of subset *S*, that is to say not all *n* are also n_s.

This restriction prevents the formation of sentences like the following:

* "Es un hombre más gordo que un hombre."

c) n_s is an element of the same class to which n_h belongs, that is to say both n_s and n_h are *n*. This restriction prevents the formation of sentences like the following:

"Es un hombre más gordo que un elefante."

Note

Our examination of the example is based on the premise that the

[3] Cf. H. Pilch, *op. cit.*

expression "más gordo que un cura" is a set phrase, a saying; thus we reject the expression "más gordo que un elefante".

The same restrictions apply to the English sentence:

"a bigger car than a Cadillac"[4]

We can write this construction:

$$\text{s-er} + n_h + \text{"than"} + n_s$$

Restriction c prevents us from saying:

* "a bigger car than a bus."

On the other hand, when we place the adjective after the noun, the second member of the comparison need not necessarily belong to the same semantic category as the first. The same applies to Spanish when we are not dealing with a set expression.

"un carro más grande que un Cadillac." / un ómnibus."

We can interpret this case as follows:

Let *V* be the set of all vehicles *v*, *A* the set of all cars v_a, and *C* the set of all Cadillac cars, v_c. The relation is transitive:

$$R = \{(v_a \in A \sim C, v_o \in C \subset A \text{ or } V \sim A) \mid v_a \text{ es más } s \text{ que } v_o\}$$

In this expression 'v_o' means an element which can belong both to the difference $V \sim A$ and to the subset $C \subset A$.

Graphically:

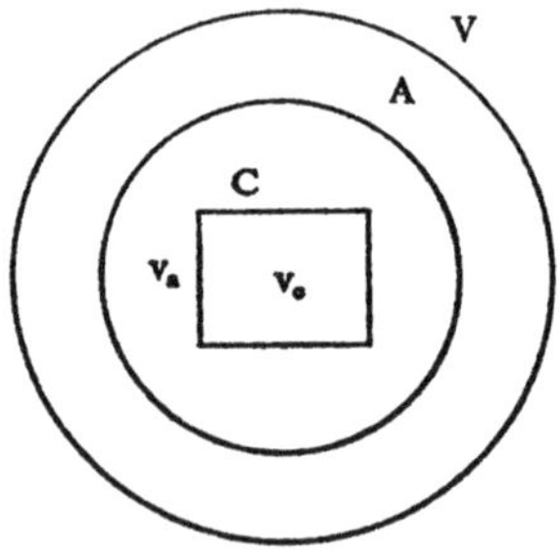

[4] See: H. Pilch, *op. cit.*

Example 46: Let us examine the endocentric construction of the Spanish sentence:

"Es una casa más cómoda que la que tiene mi amigo."

Let H be the set of all houses h and A the set of all my friend's houses a, then we have the following transitive relation:

$$R = \{(h_o \in C \sim A, h_a \in H \cap A) \mid h_o \text{ más } s \text{ que } h_a\}$$

Graphically:

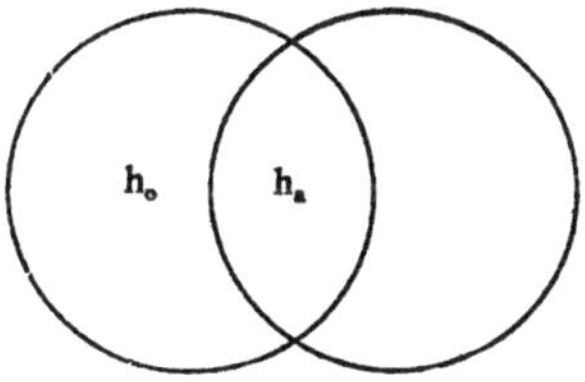

h_o is an indefinite variable in $H \sim A$.

Example 47: Let us examine the relation in the endocentric construction of the sentence

"Es un casa más grande que cómoda."

Let P be the set of the comparable adjectives p. We get the following transitive relation:

$$R = \{(p_1 \in P, p_2 \in P) \mid \text{más } p_1 \text{ que } p_2\}$$

According to this formula, the two qualities to be compared must be different from each other.

Example 48: Let us examine the relation in the exocentric construction of the sentence

"El hombre es más gordo que un cura."

This gives us the following sets:

$N = \{n \mid n \text{ is a human being}\}$
$S = \{n_s \mid n_s \text{ is a priest}\}$
$V_c = \{v_c \mid v_c \text{ is linking-verb such as "ser", "parecer", etc.}\}$

The relation is transitive and may also be written as follows:

$$R = \{(n_h, n_s) \mid n_h \in N \sim S \;\&\; n_s \in S \;\&\; v_c \text{ is a linking verb, } \& \; n_h \text{ más } s \text{ que } n_s\}$$

The following restrictions apply:

a) n_h and n_s are both elements of set N since $S \subset N$; but they are different elements since $n_h \in N \sim S$ and $n_s \in S$. Thus we cannot form sentences like the following:

* "El hombre es más gordo que un hombre."

However, a comparison of two elements of the difference $N \sim S$ is possible if we are not dealing with a set expression and if the members of the comparison show an expansion of meaning:

"Este hombre es más gordo que el que vive en la esquina."

b) Another restriction, which also applies to this sentence, is that *s* is not the nucleus of an endocentric construction and so the following construction is not possible:

* "El hombre es muy más gordo que un cura."[5]

Example 49: Let us examine the relation of the construction of this sentence:

"Hoy he comido una manzana más grande que ayer."

We have the following sets:

$$H = \{n_h \mid n_h \text{ is an apple eaten today}\}$$
$$Y = \{n_y \mid n_y \text{ is an apple eaten yesterday}\}$$

Obviously there is no intersection between these two sets. The relation is transitive:

$$R = \{(n_h, n_y)\; n_h \in H \;\&\; n_y \in Y \;\&\; n_h \text{ más } s \text{ que } n_y\}$$

This relation contains the implicit restriction that the adverbs which determine the two sets — today, yesterday — are comparable, so that the following type of construction is not possible:

* "Hoy he comido una manzana más grande que allí."

The adverbs must be from the same class, i.e. the same subset (adverbs of time) of the set of adverbs.

5 But "*mucho* más gorde que".

Example 50: Let us examine the relation of the construction of the sentence:

"María es más inteligente de lo que Juan piensa."

This sentence contains two implicit judgements: an evaluation of María's intelligence made by Juan, and one made by the person who says or thinks the sentence. If we denote the set of opinions on María's intelligence as I, we get the following transitive relation:

$$R = \{(i_a, i_j\}\ i_a \in I \ \&\ i_j \in I;\ i_a > i_j\}$$

in which i_a is a quantitive evaluation of María's intelligence made by the person who utters the sentence, and i_j the quantitive evaluation of Maria's intelligence made by Juan.

The following restrictions apply to this construction:

a) The expansion "de lo que Juan piensa" can only have as predicate certain verbs like "imagine, believe, suppose" etc. For this reason, the following type of construction is not possible:

* "María es más inteligente de lo que Juan estudia."

b) In the same construction the expansion could also be a predicate with an impersonal adjective:

"María es más inteligente de lo que sea normal."

Example 51*:* Let us examine the relation of the construction of the sentence:

"Compré un carro más grande que Juan."

This is an elliptical construction for:

"Compré un carro más grande que aquél que compró Juan."

The relation is transitive within the set "cars". The following restrictions apply:

a) Element n_1 is an undefined element of set C (cars), different from n_2.

b) In the expansion the subject of the relative clause must be a determined element; this is also true for the elliptical form; for this reason the following type of sentence is not possible without a change of meaning:

* "Compré un gatito más tierno que una niña."

This sentence states that the kitten is more tender than a little girl.

2. FURTHER EXAMPLES

Example 52: Assimilation plays a very important part in linguistic change. It affects, for example, many words in Romance languages in their passage from Latin. Let us take as an example the derivation of the following Italian words from Latin:[6]

"amministrazione"	<	"administratio"
"attivo"	<	"activus"
"assolvere"	<	"absolvere"
"settimo"	<	"septimus"

We can see that the gemination is a regressive assimilation, as the assimilated sound precedes the sound which assimilates it.

Let *A* be the set of the ordered pairs (d,m), (k,t), (b,s), (p,t), and *B* the set of the ordered pairs (m,m), (t,t), (s,s), (p,t). Set *B* is linked to set *A* by means of the operation of the regressive assimilation *a*. Since the range of the function also contains an element which is not an image of another element in its range, according to the function defined, it is an into-function or an into-mapping. Thus we cannot form an inverse function of this example since it is not an onto-mapping.

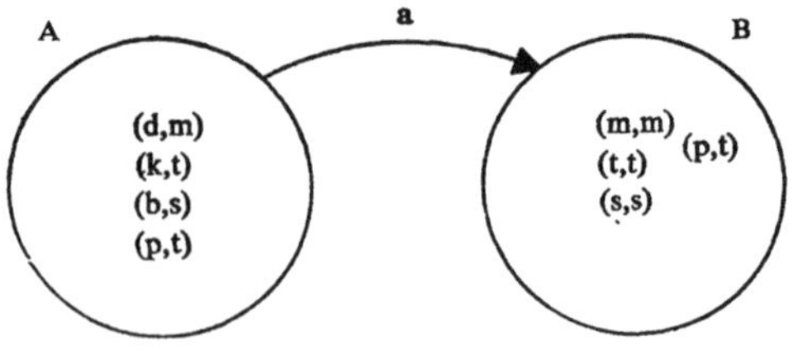

[6] But also exceptions exist: "captazione" < "captatio".

Example 53: In constructions containing a number of words of the same class the placing of the comma sometimes has a distinct semantic function. Compare the following two situations:

a) In a room there are 3 big round tables and 2 small round tables. In this situation I could give the following order:

"Please bring me the big round tables."

The comma is not used after 'big' because in the same room there are also *small* round tables.

In terms of set theory:

If R is the set of the round tables, B the set of the big round tables b, and S the set of the small round tables m, sets B and S are subsets of R:

$$R = \{B, S\}$$

$$B = \{b_1, b_2, b_3\} \qquad S = \{m_1, m_2\}$$

$$B \subset R \supset S$$

b) In a room there are 3 big round tables, and 2 small square tables. The same order as in a) must have a comma after 'big':

"Please bring me the big, round tables."

The comma — and the pause in speech — indicate that only the tables wanted are big and round; the others are neither big nor round.

In terms of set theory:

If G is the set of the big things, R the set of the round tables, S the set of the small things, and C the set of the square tables, we have:

$$R \subset G$$

$$C \subset S$$

IV

FINAL OBSERVATIONS

In the course of our study we have seen possible applications of set theory to linguistics, with a variety of examples; but at no point have we pretended to have exhausted the possibilities in this field.

Perhaps this is the point to ask what sort of linguistic problems can be illuminated by set theory. In our opinion the problems most suitable for this treatment are those concerned with structural aspects of the formal side of language, such as the phonemic and syntactical fields, but also lexico-morphological and semantic points.

One danger should be warned against: It may easily happen that, for the sake of mathematical elegance and in order to "make a scientific impression" using mathematical formulae, these methods may be applied to linguistic problems unsuitable for them. Among this group of problems are those connected with psycholinguistics, with "inneren Sprachform" within the sphere of "inhaltsbezogener Sprachforschung".

For us, linguistics is a humanistic science 'par excellence'. In the cultural manifestations of man there exist factors which cannot all be reduced to the logical values of "true" and "false", and which bring into consideration forces which distinguish man as a rational animal from a machine. Each language has its own "logic", the product of the individual development of the linguistic community which speaks it. In the field of linguistics conceived as "Geisteswissenschaft" it is interesting to consider, for example, why in one language syntactical structures based on discontinuous immediate constituents are more common than in others, why in one language syntactical structures have greater depth (in the sense of Yngve's Depth Hypothesis) than in others, why there are languages which tend to avoid a personal subject, etc. Set theory seems to us to be of little use to illuminate this type of problem, for we do not get past the descriptive stage, while the science of language, in its second phase, cannot do without the understanding and interpretation of the significance of a linguistic structure.

SELECTED BIBLIOGRAPHY*

1. GENERAL WORKS

Behnke, H., Remmert, R., Steiner, H. G., and Tietz, H. (ed.), *Mathematik*, 1 (Frankfurt am Main, Fischer Bücherei, 1964).

Haupt, D., *Mengenlehre leicht verständlich* (Leipzig, Fachbuchverlag Leipzig, 1965).

Kamke, E., *Mengenlehre* (Berlin, Walter de Gruyter, 1962).

Leung, K. T., and Chen, D. L. G., *Elementary Set Theory* (Hong Kong, Hong Kong University Press, 1964).

Lipschutz, S., *Set Theory and Related Topics* (New York, Schaum Publishing Co., 1964).

Selby, S., and Sweet, L., *Set, Relations, Functions, an Introduction* (New York, McGraw-Hill Book Co., 1963).

2. SET THEORY APPLIED TO LINGUISTICS

Abernathy, R., "The Problem of Linguistic Equivalence", in *Jakobson.*

Bach, E., *An Introduction to Transformational Grammars* (New York, Holt, Rinehart and Winston, 1964).

Bar-Hillel, Y., *Language and Information* (Reading, Addison-Wesley Publishing Co., 1964).

Chomsky, N., "Transformational Analysis", Dissertation, University of Pennsylvania, 1955.

——, *Syntactic Structures* (The Hague, Mouton & Co., 1957).

——, "On Certain Formal Properties of Grammars", in *Information and Control*, 2:137-67.

——, "On the Notion 'Rule of Grammar'", in *Jakobson.*

——, "Formal Properties of Grammars", in *Handbook of Mathematical Psychology* II, Eds.: R. D. Luce, R. R. Bush, and E. Galanter (New York, John Wiley & Sons, 1963).

——, and Miller, G. A., "Introduction to the Formal Analysis of Natural Languages", in *Handbook of Mathematical Psychology* II, Eds.: R. D.Luce, R. R. Bush, and E. Galanter (New York, John Wiley and Sons, 1963).

* Bibliographic completeness has not been aimed at.

Cooper, W. S., *Set Theory and Syntactic Description* (The Hague, Mouton & Co., 1964).

Curry, H. B., "Some Logical Aspects of Grammatical Structure", in *Jakobson.*

Hays, D. G., "On the Value of Dependency Connection", in *Proceedings of the 1961 International Conference of Languages and Applied LanguageAnalysis,* Vol. II (London, Her Majesty's Stationery Office, 1962).

Ihm, P. and Lecerf, Y., *Eléments pour une grammaire générale des langues projectives* (EURATOM, Rapport CETIS EUR 210.f, 1963).

Jakobson, R. (ed.), *Structure of Language and its Mathematical Aspects* (= *Proceedings of Symposia in Applied Mathematics,* Vol. XII) (Providence, American Mathematical Society, 1961).

Jones, K. S., "Mechanized Semantic Classification", in *Proceedings of the 1961 International Conference of Languages and Applied Language Analysis,* Vol. II (London, Her Majesty's Stationery Office, 1962).

Lambeck, J., "On the Calculus of Syntactic Types", in *Jakobson.*

Marcus, S., *Lingvistica Matematica* (Bucarest, Editura Didactica si Pedagogica, 1963) (2nd ed. 1966).

——, "Langues complètement adéquates et langues régulières", in *Zeitschrift für mathematische Logik und Grundlagen der Mathematik,* Vol. 10, pp. 7-13 (1964).

——, *Algebraic Linguistics; Analytical Models* (New York and London, Academic Press, 1967).

——, *Introduction mathématique a la linguistique structurale* (Paris, Dunod, 1967).

Mitchell, R. P., "A Note on Categorial Grammars", in *Proceedings of the 1961 International Conference of Languages and Applied Language Analysis,* Vol. I (London, Her Majesty's Stationery Office, 1962).

Mizutani, S., "On a Fundamental Assumption in Linguistics", in *Annals of the Japanese Association of Philosophy of Science,* Vol. 2, No. 3 (1963).

Mulder, J. W. F., "Some Operations with Sets in Language", in *Foundations of Language,* Vol. 1, No. 1 (1965).

Pilch, H., "Comparative Constructions in English", in *Language* 41.37 (1965).

Sestier, A., "Contribution à une théorie ensembliste des classifications linguistiques", communication au 1er Congrès de l'AFCAL, Grenoble, 1960.

Wells, R., "A Measure of Subjective Information", in *Jakobson.*

JANUA LINGUARUM

STUDIA MEMORIAE NICOLAI VAN WIJK DEDICATA

Edited by C. H. van Schooneveld

SERIES MINOR

1. ROMAN JAKOBSON and MORRIS HALLE: Fundamentals of Language. 1956. 97 pp. Gld. 6.—

3. EMIL PETROVICI: Kann das Phonemsystem einer Sprache durch fremden Einfluss umgestaltet werden? Zum slavischen Einfluss auf das rumänische Lautsystem. 1957. 44 pp. Gld. 4.—

4. NOAM CHOMSKY: Syntactic Structures. Sixth printing. 1965. 118 pp. Gld. 8.—

5. N. VAN WIJK: Die baltischen und slavischen Akzent- und Intonationssysteme: Ein Beitrag zur Erforschung der baltisch-slavischen Verwandtschaftsverhältnisse. 2nd ed. 1958. 160 pp. Gld. 15.—

8. AERT H. KUIPERS: Phoneme and Morpheme in Kabardian (Eastern Adyghe). 1960. 124 pp. Gld. 16.—

9. A. ROSETTI: Sur la théorie de la syllable. Deuxième édition, refondue et augmentée. 1963. 43 pp. Gld. 7.50

10. URIEL and BEATRICE WEINREICH: Yiddish Language and Folklore: A Selective Bibliography for Research. 1959. 66 pp. Gld. 6.—

14. DWIGHT L. BOLINGER: Generality, Gradience, and the All-or-None. 1961. 46 pp. Gld. 5.50

19. WALERIAN ŚWIECZKOWSKI: Word Order Patterning in Middle English: A quantitative study based on *Piers Plowman* and *Middle English Sermons*. 1962. 114 pp. Gld. 16.—

20. FINNGEIR HIORTH: Zur formalen Charakterisierung des Satzes. 1962. 152 pp. Gld. 15.—

22. E. F. HADEN, M. S. HAN, and Y. W. HAN: A Resonance Theory for Linguistics 1962. 51 pp. Gld. 7.—

23. SAMUEL R. LEVIN: Linguistic Structures in Poetry. Second printing. 1964. 64 pp. Gld. 8.—

24. ALPHONSE JUILLAND and JAMES MACRIS: The English Verb System. 1962. 81 pp. Gld. 8.—

25. IVAN FONÁGY: Die Metaphern in der Phonetik: Ein Beitrag zur Entwicklungsgeschichte des wissenschaftlichen Denkens 1963. 132 pp., 5 figs. Gld. 14.—

26. H. MOL: Fundamentals of Phonetics, I: The Organ of Hearing. 1963. 70 pp., 28 figs. Gld. 9.—

29. PUNYA SLOKA RAY: Language Standardization: Studies in Prescriptive Linguistics. 1963. 159 pp. Gld. 16.—

30. PAUL L. GARVIN: On Linguistic Method: Selected Papers. 1964. 158 pp. Gld. 14.—

31. LÁSZLÓ ANTAL: Content, Meaning, and Understanding. 1964. 63 pp. Gld. 8.—

32. GEORGES MOUNIN: La machine à traduire: Histoire des problèmes linguistiques. 1964. 209 pp. Gld. 22.—

33. ROBERT E. LONGACRE: Grammar Discovery Procedure: A Field Manual. 1964. 162 pp. 2nd ed. Gld. 12.—

35. LUIS J. PRIETO: Principes de noologie: Fondements de la théorie fonctionnelle du signifié. Préface d'André Martinet. 1964. 130 pp., 36 figs. Gld. 18.—

36. SEYMOUR CHATMAN: A Theory of Meter. 1965. 229 pp., many graphs, 2 plates. Gld. 21.—

37. WAYNE TOSH: Syntactic Translation. 1965. 162 pp., 58 figs. Gld. 21.—

38. NOAM CHOMSKY: Current Issues in Linguistic Theory. 1964. 119 pp. Gld. 10.—

39. D. CRYSTAL and R. QUIRK: Systems of Prosodic and Paralinguistic Features in English. 1964. 94 pp., 16 plates. Gld. 12.—

40. FERENC PAPP: Mathematical Linguistics in the Soviet Union. 1966. 165 pp. Gld. 22.—

42. MILKA IVIĆ: Trends in Linguistics. Translated by Muriel Heppell. 1965. 260 pp. Gld. 25.—

43. ISTVÁN FODOR: The Rate of Linguistic Change: Limits of the Application of Mathematical Methods on Linguistics. 1965. 85 pp., some figs. Gld. 12.—

45. WARREN H. FAY: Temporal Sequence in the Perception of Speech. 1966. 126 pp., 29 figs. Gld. 18.—

46. A. CAPELL: Studies in Socio-Linguistics. 1966. 167 pp., 2 tables. Gld. 20.—

47. BOWMAN CLARKE: Language and Natural Theology. 1966. 181 pp. Gld. 24.—

49. SAMUEL ABRAHAM and FERENC KIEFER: A Theory of Structural Semantics. 1966. 98 pp., 20 figs. Gld. 14.—

50. ROBERT J. SCHOLES: Phonotactic Grammatically. 1966. 117 pp., many figs. Gld. 14.—

65. ROGER L. BROWN: Wilhelm von Humboldt's Conception of Linguistic Relativity. 1967. 132 pp. Gld. 16.—

66. EUGENE J. BRIERE: A Psycholinguistic Study of Phonological Interference. 1968. 84 pp. Gld. 12.—

67. ROBERT L. MILLER: The Linguistic Relativity Principle and New Humboldtian Ethnolinguistics: A History and Appraisal 1968. 127 pp. Gld. 16.—

73. CHARLES HOCKETT: The State of the Art: A Critical Review of Current American Linguistic Theory. 1968. 123 pp. Gld. 14.—

MOUTON · PUBLISHERS · THE HAGUE

Bei Fragen zur Produktsicherheit wenden Sie sich bitte an:
If you have any questions regarding product safety,
please contact:

Walter de Gruyter GmbH
Genthiner Straße 13
10785 Berlin
productsafety@degruyterbrill.com